YOUR KNOWLEDGE HAS VALUE

Bibliographic information published by the German National Library:

The German National Library lists this publication in the National Bibliography; detailed bibliographic data are available on the Internet at http://dnb.dnb.de .

Imprint:

Copyright © 2017 GRIN Verlag, Open Publishing GmbH
Print and binding: Books on Demand GmbH, Norderstedt Germany
ISBN: 9783668426955

This book at GRIN:

http://www.grin.com/en/e-book/355224/microbe-supported-enhanced-production-of-rosmarinic-acid-of-medicinal-plants

Pratibha Chaturvedi

Microbe supported enhanced production of Rosmarinic acid of medicinal plants in vitro

GRIN Publishing

Enhanced production of Rosmarinic acid in *Salvia officinalis* and *Ocimum sanctum* suspension culture using entrapped *E coli* K 12 alginate beads, determined by HPTLC.

Pratibha Chaturvedi

Loyola Center for Research and Development , St Xaviers college campus, Ahemadabad 09

Abstract

In the described study , the bio-enhancement of Rosmarinic acid(RA) was done successfully in suspension culture of *Salvia officinalis* and *Ocimum sanctum* suspended with E coli K12 entrapped alginate beads.It is reported that Tyrosin Amino transferase has a key role in the production of Rosmarinic acid , which is present in *E coli* K 12. Quantification by HPTLC indicated that the maximum RA content was noted in 4 weeks old (2050 mg/l) in *Salvia officinalis* suspension culture fed with Tyrosin 10 mg/100ml followed by 4 weeks old *Ocimum sanctum* suspension culture fed with Tyrosin 10 mg/100ml(715 mg/l), that are about 18 fold and 5 folds enhancement as compared to control. The method for significantly enhancement of RA was reported for the first time.

Key words: Rosmarinic acid ; *Salvia officinalis* ;*Ocimum sanctum* suspension culture ; *E coli* K 12 , Tyrosin Amino Transferase, HPTLC

Contents

Introduction

Rosmarinic acid (RA) is a water-soluble phenolic acid , which is mainly found in the plant species of Lamiaceae and Boraginaceae (1). RA derived from caffeic acid and (*R*)-(+)-3-(3, 4-dihydroxyphenyl) lactic acid represents one of the most common caffeic esters in plant material, was originally identified in *Rosmarinus officinalis* L. and accumulated constitutively (2). .In addition to many essential nutritional components, plants contain phenolic substances, agroup of biologically active non-nutrients RA is one of them .RA possesses various biological activities such as antimicrobial, anti-inflammatory, antimutagenic, improvement of cognitive performance, prevention of the development of Alzheimer's disease, cardioprotective effects, reduction of the severity of kidney diseases, antioxidant and cancer chemoprevention (3, 4). *Salvia officinalis* and *Ocimum sanctum*(Lamiaceae) have been used extensively in various fields, such as medicine, food industries, as well asperfume and cosmetics productions (Adinee, *et al.*, 2008). *Salvia officinalis* and *Ocimum sanctum* are reported to contain Rosmarinic acid (5, 12).It is known that most of the medicinal effects of these plants are related to its active ingredient, (Park *et al.*, 2008). The production of secondary metabolites through a cell culture technology of renowned medicinal plants has been a challenging subject for many researchers. It has been established that cell cultures, obtained from plants of Boraginaceae family such as *Lithospermum erythrorhizon* and *Anchusa officinalis*, produce considerable amounts of rosmarinic acid (5, 6). Nasiri et al., 2014 have reported the role of Tyrosine amino transferase(TAT) in the production of Rosmarinic acid in *Melissa officinalis*.. TAT catalyses the reversible transfer of an amino group between L-tyrosine and L-glutamate using a-ketoglutarate and 4-hydroxyphenylpyruvate as co-substrates. Aminotransferases in general play an important role in nitrogen metabolism and are said to be responsible for the synthesis, in fungi, of the so-called secondary amino acids from the primary amino acids (Echetebu 1982). Tyrosine aminotransferase (TyrB) known as aromatic-amino acid aminotransferase, is a broad-specificity enzyme that catalyzes the final step in tyrosine, leucine, and phenylalanine biosynthesis. Since this important enzyme is present in *Escherichia.coli* K 12 strain (Keseler et al.,2011). Hence in the present study, *Escherichia coli* K 12 was used in immobilized form to enhance the Rosmarinic acid production in *Salvia officinalis* and *Ocimum sanctum* suspension culture as TAT is known to has extracellular nature.. The used technology is not well documented.

Material and Methods

Plant material

The leaves of *Salvia officinalis* and *Ocimum sanctum* collected from green house of Loyola Center for Research and Development, St. Xavier's college campus. The leaves were washed with teepol , rinsed and subjected for Bovistein (1%) treatment for 15 minutes. After washing in running water, these were used for further experiment in aseptic condition.

Suspension media

Zenk production media(Zenk 1975) with 5% sucrose incorporated with Tyrosin (7.5 and 10 mg/100 ml), Jasmonic acid (20, 40 mg/ 100 ml), Caffiec acid (20 and 40 mg/ 100 ml) separately were prepared and autoclaved at 121 ^{o} c temperature . Leaves were surface sterilized with 0.1% of mercuric chloride for 3 min. followed by sterilized distilled water separately in aseptic condition. Sterilized leaves(1 g) of both of plant spp were macerated in laminar flow and inoculated in to sterilized Zenk media (100 ml) supplemented with Tyrosin (7.5 and 10 mg/100 ml), Jasmonic acid (20, 40 mg/ 100 ml), Caffiec acid (20 and 40 mg/ 100 ml) separately. pH was 5.6-5.8 before autoclaving.

Preparation of beads of *E coli* K12

E coli K12 strain was bought from Microbial Type Culture Collection and Gene Bank Institute of Microbial Technology (MTCC), Chandigarh. The lyophilized culture were revived in L B liquid media. 24 hrss old culture of *E coli* K 12 was taken (20 µl /20 ml) and centrifuged at 8000rpm. Palate was suspended in 1 ml of N broth in aseptic condition and was added to 3.6% Sodium alginate.

Beads were prepared by adding this in 4% cold $CaCl_2$.These prepared beads were washed with sterile distilled water and inoculated in suspension production media of *Salvia officinalis* and *Ocimum sanctum* in aseptic condition. These inoculated media were incubated on rotary shaker at 110 rpm at 25 ^{o}c and 70 % humidity with 16 hrs of photoperiod and 8 hrs of dark period. Samples were harvested at the time interval of 2, 3 and 4 weeks and extraction for Rosmarinic acid were carried our separately.

Down stream process for Rosmarinic acid extraction

The harvested liquid media for both of the plant spp were filtered by using what man paper no 1 and filtrate were dried on hot plate at 40 ^{0}c temperature till thick slurry. All the resultant samples were kept in desiccator for further dryness for 24 hrs and put them in -20 deep fridge with 5 ml of 70% ethanol for 36 hrs in each sample. The supernatant of all the samples were pipette out and subjected for dryness at room temperature separately. The dried , weighed various samples of 2, 3 and 4 weeks age of both the plant spp (*S. officinalis* and *O. sanctum*)were then subjected for High Performance Thin Layer Chromatographic (HPTLC) analysis and FTIR spectral studies with standard compound of Rosmarinic acid (purchased from Sigma Alderich)

High Performance Thin Layer Chromatography (HPTLC) Analysis

HPTLC analysis (Camag software controlled HPTLC System) of various samples were done with standard compound of Rosmarinic acid. Toluene: Ethyl acetate (9:1) was used as mobile phase. The spot corresponding with standard compound of Rosmarinic acid was noted and scanning was done at 285 nm. Rf value was 0.1 was observed. Detection was also done by spraying developed plates with 5%ethnolic ferric chloride. Dark blue color was observed corresponding with standard Rosmarinic acid. The quantitative estimation of Rosmarinic acid in all samples were calculated by using a calibration curve.

Infra Red Spectral Studies

Infra Red Spectral studies of all used samples were carried out with standard compound of Rosmarinic acid . (By using equipment Buck scientific 500).

Statistical analysis

Statistical analysis of obtained Data was done by using t test of Graph ped prism software .

Results and Discussion

Plant Tissue culture has been served as a mean for secondary metabolites production. The strong and growing demand in today's marketplace has refocused attention on *in vitro* plant materials as potential factories for secondary phytochemical products.   Chaturvedi et al., 2009, 2010, 2011, 2013, 2014, 2015 , 2016 have extensively studied the enhancement of secondary metabolites by using different compound in tissue culture of medicinal plants. The treatment of plant cells with biotic and/or abiotic elicitors has been a useful strategy to enhance secondary metabolite production in cell cultures(Karuppusamy et al., 2009). Tyr B gene of *E. coli* strain K-12 govern the production of Tyrosin Amino Transferase (TAT) . TAT,   a broad-specificity enzyme, which catalyzes the transamination of 2-ketoisocaproate, p-hydroxy phenyl pyruvate, and phenyl pyruvate to yield leucine, tyrosine and phenylalanine, respectively . (Huang, 2009).

Hence keeping all the facts in view,in the present study various elicitors were used to bio-enhance the Rosmarinic acid(RA)by using tissue culture technique. The production of RA, a caffiec acid ester compound of medicinal value has been successfully induced by using alginate beads of *E coli* K12 bacterium in suspension culture of *Salvia officinalis* and *Ocimum sanctum* separately supplemented with various elicitors. Enhancement as observed in all the samples but the maximum content was noted in 4 weeks old (2050 mg/l) in *Salvia officinalis* suspension culture fed with Tyrosin 10 mg/100ml followed by 4 weeks old *Ocimum sanctum* suspension culture fed with Tyrosin 10 mg/100ml(715 mg/l), that are about 18 fold and 5 folds enhancement as compared to control (Table 1,2)..It is already reported. That

this extra cellular enzyme catalyzes the first step of the tyrosine pathway leading to the formation of rosmarinic acid (alpha-O-caffeoyl-3,4-dihydroxyphenyllactic acid) in suspension cultures of *Anchusa officinalis* L(De-Eknamkul et al., 1887) .This study supports the enhanced production of Rosmarinic acid in our present findings. The another reason for induction of RA content in 4 weeks old culture of both of the plant spp is may be due to given stress to E coli K 12 bacterium by immobilization, that lead to enhance the production of Tyrosin . It is well reviewed that Tyrosin has a key role in production of Rosmarinic acid as precursor compound in Biosynthetic pathway.(Fig 3.).The main advantage of the developed technology is the use of natural resource for the overproduction of RA. E coli (k-12) can be multiplied into number of copies in short time period in laboratory condition, which made the developed technology as economically feasible and affordable source for RA production.

S No	Treatment(mg/100ml)	Rosmarinic acid content in *Salvia officinalis* in mg/l		
		2 week	3 week	4 week
1	JA 40	424±4.11	457±4.33	483±6.22
2	JA20	211±7.71	340±3.87	627±6.78
3	Tyrosin10	487±3.92	715±6.66	2050±3.69
4	Tyrosin 7.5	300±6.61	696±8.91	1440±1.43
5	Caffiec acid 40	235±8.11	402±8.73	1180±6.12
6	Caffiec acid20	389±6.81	507±5.88	685±4.32
7	Control (without beads)	124±5.71	175±8.11	189±3.22
8	Control(Leaves)	115±7.88		

Table1. Bio-enhancement of Rosmarinic acid in *Salvia officinalis* using E coli K12 alaginate beads. Mean ± SD of three replicates p value <0.0001

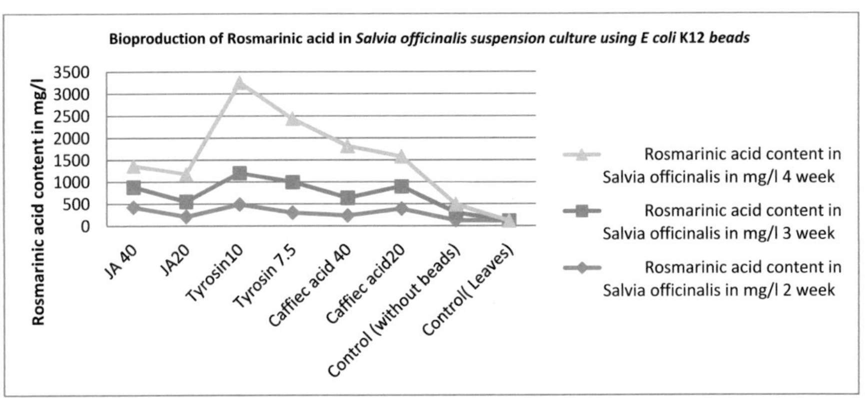

Fig1. Bio-enhancement of Rosmarinic acid in *Salvia officinalis* using E coli K12 alaginate beads

SNo	Treatment (mg/100 ml)	Rosmarinic acid content in *Ocimum sanctum* in mg/l		
		2weeks	3 weeks	4weeks
1.	JA 40	281±3.11	424±4.13	457±4.21
2.	JA20	155±5.75	211±6.12	340±4.88
3.	Tyrosin10	251±7.13	487±7.13	715±8.47
4.	Tyrosin 7.5	301±5.13	320±4.26	696±9.01
5.	Caffiec acid 40	256±9.11	285±3.88	402±9.10
6.	Caffiec acid20	153±0.99	389±5.97	507±3.88
7.	Control ((without beads)	146±1.32	154±3.76	175±2.74
8.	Control(Leaves)	145±1.99		

Table 1. Bio-enhancement of Rosmarinic acid in *Ocimum sanctum* using E coli K12 alginate beads. Mean± SD of three replicates p value <0.0001

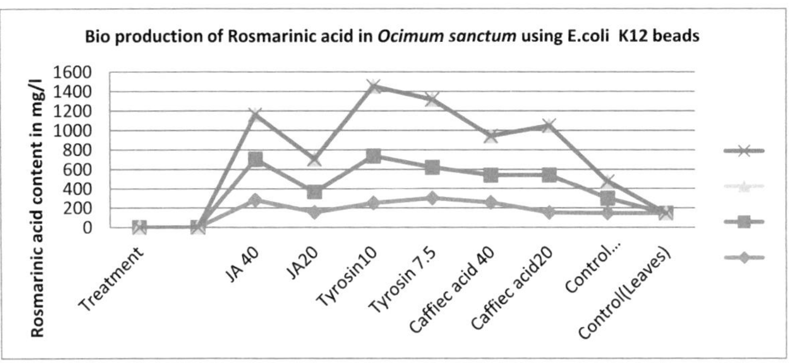

Fig. 2. Bio-enhancement of Rosmarinic acid in *Ocimum sanctum* using E coli K12 alaginate beads

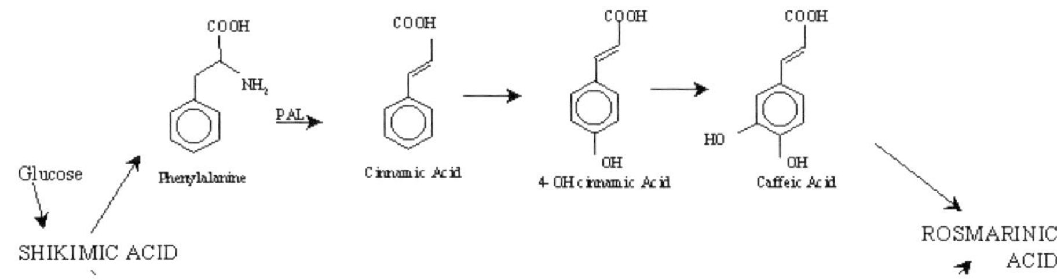

Fig 3 Proposed biosynthetic pathway of Rosmarinic acid (Shetty 1997)

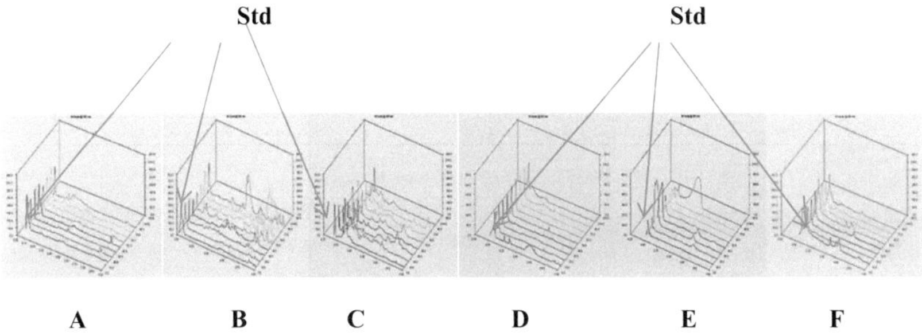

Fig.4 HPTLC analysis of *Salvia officinalis* 2,3,4 wks suspension culture (A,B,C) and *Ocimum sanctum* 2,3,4 wks suspension culture suspended with *E coli* K12 strain entrapped alaginate beads with standard compound of Rosmarinic acid.

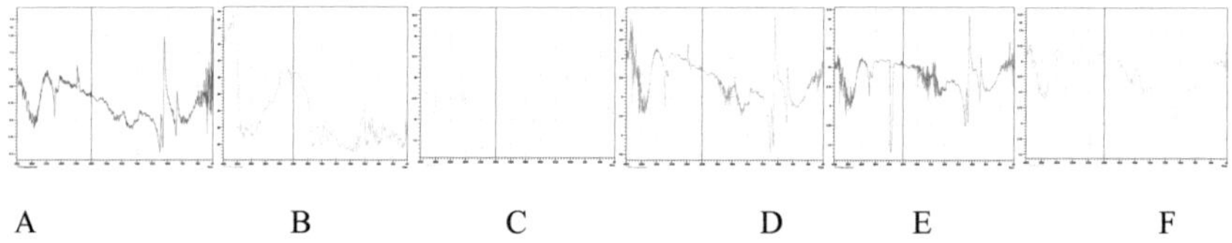

A   B   C   D   E   F

Fig.5. FTIR spectral studies of samples of four weeks old suspension culture of *Salvia officinalis* **and** *Ocimum sanctum* **A std   ; B.** *Salvia officinalis* **suspension culture fed with Tyrosin 7.5 mg/100 ml ; C.** *Ocimum sanctum* **suspension culture fed with Tyro 7.5 mg/100 ml ; D.** *Salvia officinalis* **suspension culture fed with Tyrosin 10 mg/100 ml ; E.** *Ocimum sanctum* **fed with Tyrosin 10mg/100 ml ; F. control**

S No	Peak at wavelengt (nm)	Group Denoted
1.	1750	Ester
2.	980	Carboxillic acid
3.	1598	Aromatic

Table 3.Infrared spectral studies showing the characteristic peaks of treated samples,corresponded with standard compound of Rosmarinic acid

Conclusion

The production of Rosmarinic acid was observed significantly enhance d in four weeks old suspension culture of Salvia officinalis(18 folds) and Ocimum sanctum(five folds) fed with Tyrosin 10 mg/100 ml agitated with Ecoli K12 beads. The Significance of the study embedded in the fact that the developed technology will be useful for industrial production of Rosmarinic acid, a important medicinally important compound.

Acknowledgement

Authors are grateful to Xavier's Research Foundation for providing financial assistance and Miss Disha and Miss Kumud for FTIR spectral studies.

References

Abu-shanab, B., Adwan, G., Jarrar, N. and Adwan, K. 2006. Antibacterial Activity ofFour Plant Extracts Used in Palestine in Folkloric Medicine against Methicillin-Resistant. *Staphylococcus aureus. Turk. J.Biol.*, **30**: 195-198.

Adinee, J., Piri, Kh. and Karami, O. 2008.Essential Oil Component in Flower of LemonBalm (*Melissa officinalis* L.). *Am. J. Biochem.Biotechol.*, **4**: 277-278.

Park, S., Romij, M. d., Xu, H., Kyoung Kim,Y. and Young Lee, S. 2008. BiotechnologicalApplications for Rosmarinic Acid Production in Plant. *Afr. J. Biotechnol.*, 7: 4959-4965.

Molgaard P and Ravn H. Evolutionary aspects of caffeoyl esters distribution in dicotyledons. *Phytochem.* (1988) 27: 2411-2421.

Ellis BE and Towers GHN. Biogenesis of rosmarinic acid in *Mentha. Biochem. J.* (1970) 118: 291-297.

Petersen M and Simmonds MSJ. Rosmarinic acid. *Phytochem.* (2003) 62: 121-125.

Bulgakov VP, Inyushkina YV and Fedoreyev SA. Rosmarinic acid and its derivatives: biotechnology and applications. *Crit. Rev. Biotechnol.* (2012) 32: 203- 217 .

5.Hippolyte I, Marin B, Baccou JC, and Jonard R. Growth and rosmarinic acid production in cell suspension cultures of Salvia officinalis L. PlantCell Rep.1992; 11: 109-112.

[6] De-Enamkul W and Ellis BE. Rosmarinic acid production and growth characteristics of Anchusa officinalis cell suspension cultures. Planta Med.1984; 50: 346-350.

[7] Petersen M, Hausler E, Meinhard J, Karwatzki B, and Gertlowski C. Thebiosynthesis of rosmarinic acid in suspension cultures of Coleus blumei. Plant Cell Tissue Org Cult.1994; 38: 171-179.

9.Kirtikar KR, Basu BD. Indian medicinal plants. Delhi: Periodical expertsbook agency, 1993.

10.Tapan K, Maity Subhash C, Mandel BP, and Saha Pal M. Effect of Ocimumsanctum roots extracts on swimming performance in mice. PhytotherRes.2000; 14: 120–121.

[11] Singh S and Majumdar DK. Evaluation of anti-inflammatory activity offatty acids of *Ocimum sanctum* fixed oil. Ind J Exp Biol.1997; 35:380–383.

[12] Nakatani N. Antioxidant from spices and herbs. In Shahidi F (ed.),Natural antioxidants; Chemistry, health effects and applications.Champaign I: AOCS press, 1997.

5 Misawa M. *Plant tissue culture: an alternative forproduction of useful metabolites.* Daya Publishing
House, Dehli (1997) 60-62

6 Yamamoto H, Inoue K and Yazaki K. Caffeic acidoligomers in *Lithospermum erythrorhizon* cell suspension cultures. *Phytochemistry* (2000) 53: 651-657.

M. A. Nasiri-Bezenjani1, A. Riahi-Madvar2*, A. Baghizadeh2, and A. R. Ahmadi   Rosmarinic Acid Production and Expression of Tyrosine Aminotransferase Gene in *Melissa officinalis* Seedlings
in Response to Yeast Extract *J. Agr. Sci. Tech. (2014) Vol. 16: 921-930 ,921.*

Echetebu C O .Some Properties of Tyrosine Aminotransferase from *Tuichuderma viride Journal of General Microbiology* (1982), 128, 2735-2738. *Printed in Great Britain* 2735.

Keseler IM[1], Collado-Vides J, Santos-Zavaleta A, Peralta-Gil M, Gama-Castro S, Muñiz-Rascado L, Bonavides-Martinez C, Paley S, Krummenacker M, Altman T, Kaipa P, Spaulding A, Pacheco J, Latendresse M, Fulcher C, Sarker M, Shearer AG, Mackie A, Paulsen I, Gunsalus RP, Karp PD. EcoCyc: a comprehensive database of Escherichia coli biology. Nucleic Acids Res. 2011 Jan;39.

Zenk MH, el-Shagi H, Schulte U. Anthraquinone production by cell suspensioncultures of *Morinda citrifolia*. Planta Med 1975;Suppl 75:79-81.

De-Eknamkul W1, Ellis BE.Purification and characterization of tyrosine aminotransferase activities from Anchusa officinalis cell cultures. Arch Biochem Biophys. 1987 Sep;257(2):430-8.

Pratibha Chaturvedi and Abhay Chowdhary 2011. Enhancement of tylophorin in *Tylophora indica* callus J. of Phytological Research 24,1, 9-13. **Impact Factor 0.306**

Pratibha Chaturvedi and Abhay Chowdhary 2011. Strategies for enhancement of tylophorin in *Tylophora indica in vitro* . J.of Phytological Research 24,2;191-195.Impact Factor 0.376

Pratibha Chaturvedi and Abhay Chowhary 2012. Effect of different sugars on stigmasterol production in callus of *Tylophora indica* .International J of Pharmacology and Phytochemistry 2,226-228.**Impact factor IC value 5.09.**

Pratibha Chaturvedi,Sayali Sawant and Abhay Chowdhary. Variation in phyto chemical profile of *Tylophora indica* plants collected from different regions of India. J.of Phytological Res.25(1)2012. **Impact Factor 0.306.**

Pratibha Chaturvedi, Pushpa khanna and Abhay Chowdhary. 2013. Phytosterols from tissue culture of *Allium cepa* and *Trachyspermum ammi* .Journal of Pharmacognosy and Phytochemistry 1,6,43-48.

Pratibha Chaturvedi and Abhay Chowdhary .2013. Enhancement of antioxidant compound in *Tylophora indica* callus. Advances in applied science 4(2),325-330.

Pratibha Chaturvedi, Pushpa Khanna and Abhay Chowdhary 2013 Isolation, Identification and Characterization of rotenoids from *Cajanus cajan* seeds. Indian Drugs 50(18)**.(Award Winning Best Research Paper by Indian Drugs, IDMA in 2013 Natural Products Category) Impact factor 0.327**

Chaturvedi P., Chowdhary A., 2014. A novel method for bio -enhancement of anti - cancerous compound curcumin *in vitro* tissue culture of *Curcuma longa* (Apiaceae). The Journal of Bioprocess Technology. 99, 389-395.**Impact Index 3.7**

Chaturvedi P., Chowdhary A., 2014. Stigmasterol enhancement in *Tylophora indica* (Asclepeadaceae) callus. The Journal of Bioprocess Technology. 99,344-349.**Impact Index 3.7.**

Chaturvedi P.,Chowdhary A., 2014 Variation in Stigmasterol content in Medhya rasayan plant (Centella assiatica) collected from different regions. Indian drugs, March .**Impact factor 0.327.**

Pratibha Chaturvedi, Srikant Gajbhiye, Soumen Roy, Rohan Dudhale, Abhay Chowdhary. 2014. Determination of Kaempferol in extracts of Fusarium chlamydosporum, an endophytic fungi of *Tylophora indica* (Asclepeadaceae) and its anti-microbial activity IOSR Journal of Pharmacy and Biological Sciences (IOSR-JPBS) e-ISSN: 2278-3008, p-ISSN:2319-7676. Volume 9, Issue 1 Ver. V (Feb. 2014), PP 51-55 **Impact factor 1.3.**

Pratibha Chaturvedi, Shrudda Surve, Swetha Soundar, Kreena Parekh, Samit Lokhande, Abhay Chowdhary Kaempferol and Stigmasterol Content Variations in Diverse Cowpea (*Vigna unguiculata*) Seed Varieties VRI Phytomedicine, Volume 2, Issue 2, March 2014.

Chaturvedi P, Surve S, Mukherjee S, Rajopadhye S, Kasarpalkar N, Marita R, Chowdhary A. Aspartic protease gene expression and kaempferol production at different germinating stages of cow pea (*Vigna* unguiculata L.) seeds varieties European Journal of Experimental Biology, 2013, 3:126-133

Roy S, Chaurvedi P, Chowdhary A. Evaluation of anti-viral activity of essential oil of *Trachyspermum ammi* against Japanese encephalitis virus. Pharmcog Res 2015;7:263-7.pubmed index impact factor-2.04

Chavan R, Chaturvedi P and Chowdhary A: Anti-Influenza Potential of Alkaloidal Molecules of Jatropha Curcas Leaves. Int J Pharm Sci Res 2015; 6(11): 4705-11

Pratibha chaturvedi and Vincent Briganza Enhanced synthesis of Curculigoside by Stress and amino acids in static culture of *Curculigo ochioides* Gaertn (Kali Musli). Pharmacognosy Research July - September 2016 | Volume 8 | Issue 3, 193-200

Karuppusamy S. A review on trends in production of secondary metabolites from higher plants by *in vitro* tissue, organ and cell cultures. J Med Plants Res. 2009;3:1222–39.

Huang Y.T., Lyu S.Y., Chuang P.H., Hsu N.S., Li Y.S., Chan H.C., Huang C.J., Liu Y.C., Wu C.J., Yang W.B., Li T.L. In vitro characterization of enzymes involved in the synthesis of nonproteinogenic residue (2S,3S)-beta-methylphenylalanine in glycopeptide antibiotic mannopeptimycin."ChemBioChem 10:2480-2487(2009) .

Kalidas Shetty.Biotechnology to harness the benefits of dietary phenolics; focus on *Lamiaceae Asia Pacific J Clin Nutr (1997) 6(3): 162-171*